Student Guide

THOMSON

BROOKS/COLE

Australia • Canada • Mexico • Singapore • Spain • United Kingdom • United States

THOMSON
™
BROOKS/COLE

President: Michael Johnson
Acquisitions Editors: Jennifer Huber, Bob Pirtle
Assistant Editors: Kirsten Markson, Rachael Sturgeon
Editorial Assistant: Carrie Dodson
Senior Project Manager: Julia Pluss
Advertising Project Manager: Margaret Parks
Project Manager, Editorial Production: Jennifer Risden
Print/Media Buyer: Doreen Suruki
Production Service: Delgado Design, Inc.
Copy Editor: Linda Stern
Cover Image: Getty Images
Compositor: Delgado Design, Inc.
Text and Cover Printer: Transcontinental Printing/Interglobe

Printed in Canada
4 5 6 7 07 06 05 04

For more information about our products, contact us at:
Thomson Learning Academic Resource Center
1-800-423-0563

For permission to use material from this text or product, submit a request online at
http://www.thomsonrights.com.
Any additional questions about permissions can be submitted by e-mail to **thomsonrights@thomson.com.**

ISBN 0-534-39914-2

Thomson Brooks/Cole
10 Davis Drive
Belmont, CA 94002
USA

Asia
Thomson Learning
5 Shenton Way #01-01
UIC Building
Singapore 068808

Australia/New Zealand
Thomson Learning
102 Dodds Street
Southbank, Victoria 3006
Australia

Canada
Nelson
1120 Birchmount Road
Toronto, Ontario M1K 5G4
Canada

Europe/Middle East/Africa
Thomson Learning
High Holborn House
50/51 Bedford Row
London WC1R 4LR
United Kingdom

Latin America
Thomson Learning
Seneca, 53
Colonia Polanco
11560 Mexico D.F.
Mexico

Spain/Portugal
Paraninfo
Calle/Magallanes, 25
28015 Madrid, Spain

Contents

Improve Your Grade with iLrn

Welcome to the iLrn Student Guide. On the pages that follow, you will see how iLrn, an integrated, online learning system will provide you with anywhere, anytime access to tutorials and tools that will help you master the concepts of your course and be better prepared for tests, midterms, and finals.

iLrn's online environment allows you to study at your own pace (or at a pace your instructor assigns), quiz yourself, monitor your own progress, and turn in your homework and exams online. For most Brooks/Cole texts, the tutorial section provides the entire text, section by section, in PDF (portable document format) to support your online work. As you work through hands-on practice, quiz, and homework material, you are assisted by explanations, examples, and problems from the text. iLrn gives you an unsurpassed integrated learning environment, combining all the best elements of textbook learning aids, interactive learning, unlimited drill and practice, and conceptual problems to comprehensively assist your understanding.

In addition, if your professor has set up a course in iLrn, you can see all your grades and e-mail your instructor from within the program.

Be sure to take a look at the iLrn Quick Start Guide on the next page. That's the quickest way to get up and running on iLrn. And be sure to take a look at the various appendixes, where you will find additional information on using iLrn in mathematics and statistics courses.

And remember, help is just a click away at iLrn!

Quick Start Guide

Welcome to iLrn, where self-paced tutorials, online help, step-by-step exercises, assignment sheets, and plenty of tools enable you and your instructor to work as partners.

If you have never used iLrn before, see **First Time User Registration.** If you are a returning iLrn user, go to **Login**. Once you are logged in, you can **Register Additional Content or Courses** from the Main Menu.

Note: For troubleshooting help during system setup and registration, see the **Student Support > FAQs** online at **www.ilrn-support.com**.

First Time User Registration

Open your web browser and go to **http://www.ilrn.com**. This is the iLrn "Front Porch" page.

1 Click **First Time User.**

2 Follow the on-screen instructions to select your school, if necessary.

3 Enter the access code for your iLrn content (book/CD) or course, and your valid e-mail address. Click **Submit**.

4 Follow the on-screen instructions to enter a valid password and your contact information.

5 Click **Register and Enter iLrn** to finish registration and log in.

Notes:

• Your e-mail address will be used by the system as your iLrn login.

• Keep your password private and in a safe place. You will need it each time you log in.

Login

Once you have registered for iLrn, you can log in directly.

1 Go to **www.ilrn.com**.

2 Click **Login**.

3 Enter your iLrn login and password. Click **Login**.

Welcome to iLrn!

Registering New Content or Courses

Once you have registered for iLrn, it's easy to register for additional content (products) or courses.

1 **Login** to iLrn.

2 From **Main Menu > Other**, select **Enter Content Access Code** (to register a new book or CD) or **Enter Course Access Code** (to enter an iLrn course enrollment code).

3 Follow the on-screen instructions to enter the appropriate code.

Online Help

For online help with registration, click the **Student Support** link.

Click on **FAQs** (Frequently Asked Questions) for information on browser setup, Java, enrollment, and other topics. Once you log in, see **Help** at the upper right of the screen.

Student Technical Support

Phone: 800-423-0563
Monday–Friday, 8:30 A.M.–6:00 P.M. Eastern time
E-mail: tl.support@thomson.com

Navigating iLrn

The iLrn Main Menu gives you access to all of iLrn's great tools to help you learn:

General

- **My Assignments.** Find a chart of your assignments with due dates, scores, and more.
- **Progress.** Keep your own personal gradebook.
- **Syllabus.** Find information about your instructor and class.

Products

- **Courseware.** Access interactive electronic text or courseware.
- **Tutorials.** Study activities vary by book, but frequently include book explanations, exercises, quizzes, video tutorials, live online tutoring, chapter tests, and explorations.

Other

- **Change Account Information.** Change your password and other account information.
- **Enter Course Access Code.** Access the coursework and tutorials for a second course.
- **Enter Content Access Code.** Access the full set of tutorials and courseware for an iLrn textbook or supplemental product.

When you need help, use the following links, in the upper right corner of the screen.

- **FeedBack.** Let us know how iLrn is working for you.
- **Help.** Get the answers you need with context-sensitive help.
- **Resources.** Find glossaries and additional tools for your discipline.
- **Technical Support.** Go to the iLrn Technical Support Website for FAQs (frequently asked questions), student guides, new products, downloads, Report a Bug, feedback forms, and JRE (Java Runtime Environment) instructions.

Assignments and Tutorials

iLrn puts you in charge of your learning. With self-paced tutorials, detailed explanations, step-by-step examples, and quizzes, you can get the most out of your textbook—and turn in your assignments from your own room!

Assignments

Do homework assignments and tests online by clicking on **My Assignments** under "General" in the menu at the left of the screen. You'll find a chart of all your assignments with due dates, scores, notes from the instructor, the number of times you're permitted to take an assignment, extra credit, and more.

Click the **Take** button next to any assignment to start. When you complete a homework assignment or test, you'll see a chart of your results that will indicate which problems you answered correctly. Use this as a guide for further study. Meanwhile, your score on the assignment will be recorded both in the instructor's gradebook and in your own progress chart, which you can find by clicking on **Progress** in the menu at the left.

Tutorials

Study the topics most important to you by clicking on **Tutorials** under "Products" in the left menu. Select your textbook; then click on pull-down menus to choose a chapter and section. You'll find some or all of these choices:

- **Read the Book.** Find the key passages from the book to concisely define and clarify key concepts, pulled out for your context-sensitive use as you study.
- **Exercises.** Try as many of iLrn's interactive problems as you need to; an unlimited number are available, and many include step-by-step examples that walk you through solving the problem.
- **Quizzes.** Test how much you've learned using iLrn!
- **Video.** SkillBuilder sessions use video clips to present problems and demonstrate complete solutions.
- **vMentor Tutoring.** Click to enter a vMentor classroom where live instructors assist you with problems and homework.
- **Chapter Tests.** Take a 25-question test to demonstrate your mastery of a chapter.
- **Explorations.** Practice the skills for each chapter by solving these real-life problems.

Logging In to iLrn

If you have never registered or used iLrn before, see "First Time User Registration," below. If you are a returning iLrn user, go to "Login," page 7.

First Time User Registration

Registering your iLrn product or course takes just a few steps. You will need to select your school, enter your access code and e-mail address, select a password, and enter your contact information.

The iLrn Front Porch

1️⃣ Open your web browser and go to **http://www.ilrn.com**.
This is the iLrn Front Porch.

2️⃣ Click **First Time Users**.

Find Your School

If you are new to iLrn, the "Find Your School" page will open so you can select your school.

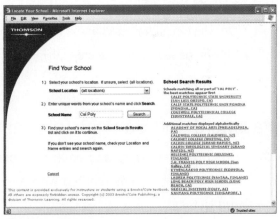

Find Your School Page

1️⃣ In the **School Location** box, select your school's location from the drop-down list. (If you're unsure or can't find it, select "all locations.")

2️⃣ In the **School Name** box, type key words of your school's name and click **Search**.
A list of schools matching your search terms will appear on the right.

3️⃣ Find your school's name on the list, and click on it to continue.

If your school is not listed at first, refine your search terms and click **Search** again. Double-check the location, and enter only key, unique parts of your school's name.

Note: You may need to find your school again later if you change computers, change schools, clear your browser "cookies," or reinstall iLrn. Use the **Find Your School** link on the "Login" page.

Access Code and E-mail

Enter Access Code and E-mail

1 In the **Access Code** box, enter the Content Access Code on the card from your book or CD, or the Course Access Code from your instructor. For some products, you'll enter the ISBN number from the back cover.

2 In the **E-mail** box, enter your e-mail address. The system will use this e-mail as your login.

Note: The e-mail address can be no longer than 30 characters. It can contain letters, numbers, period . , underscore _ , hyphen - , and @ symbol only.

3 Click **Submit**.

Password and Contact Information

Create your iLrn password and enter your contact information by following the on-screen instructions. Fields marked with an asterisk must be filled in.

Password Notes:

• Your password must be 6–15 characters long and include letters, numbers, and underscore "_" only.

• The password is "case-sensitive," meaning you must capitalize it the same way every time. For example, "BigKat" is not the same password as "bigkat" or "BIGKat."

- Be sure to choose a password that is easy to remember, but not so easy that someone else can guess it. Keep your account information private and secure.

If an entry is not accepted, correct the item as directed and enter it again.

Click **Register and Enter iLrn** to finish registration and log in.

Contact Information Page

Login

If you already have an iLrn account or have registered previously for another Thomson online product, you can log in directly.

1. Go to **www.ilrn.com** to open the "iLrn Front Porch" page.

2. Click **Login**.

3. Enter your iLrn login and password.

4. Click **Login**.

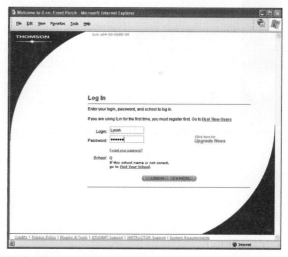

Login Page

Note: You may be asked to replace your current iLrn login with your e-mail address and to select a password (6–15 characters long), if you haven't done so already. Follow the on-screen instructions to verify and update your login, password, and contact information as needed.

Getting Started

You are now ready to get started with iLrn. Note the menu that runs along the left side of the screen. This menu, which is always evident, provides access to all the course areas, including "My Assignments," "Progress," and "Tutorials," as well as opportunities to change your password or register for an additional course or product.

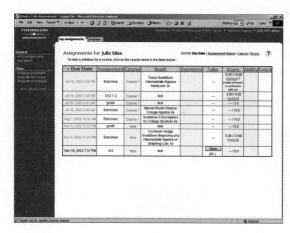

My Assignments Page

The "My Assignments" Area

The "My Assignments" page provides a chart (or syllabus) that lists the assignments and allows access to them. In this chart, you can see, for any assignment:

- Due date and time
- Course name
- Comments from the instructor (if any)
- **Take** button and number of takes allowed and completed

- Assignment name
- Book (test or text for the tutorial)
- Scores
- Extra credit (if any)

Check the chart regularly to keep up with the due dates and times of assignments.

The "Take" column includes a **Take** button to get started on each assignment (see "Access Assigned Homework" on the next page). The "Take" column also identifies how many attempts were made and how many attempts are allowed on each assignment. For instance, "0/--)" indicates that you have not attempted the assignment and that you may attempt it an unlimited number of times. Similarly, "(0/1)" indicates that you have not attempted the assignment and that you may attempt it only once.

Access Assigned Homework

If your instructor has assigned tutorials or homework problems, you will see those listed in the chart on the "My Assignments" page. Click on the **Take** button in the chart to get started on an assignment. Any work completed in the assigned tutorial is graded and recorded in the instructor's gradebook.

Access Assigned Quizzes and Tests

If your instructor has assigned quizzes or tests, then you will see those listed in the chart on the "My Assignments" page.

1 To get started, click on the assignment's **Take** button.

Note: Do not click on **Take** until you are ready to attempt the assignment. Clicking on **Take** but not finishing a quiz or test can lower your score.

2 On the screen that appears, click on **Click Here.** The assignment will open.

3 The screen for your assignment is organized as follows:

- **Previous and Next Tabs.** By clicking on these tabs at the top of the page, you can navigate forward and backward between questions.

- **Jump to Question.** By choosing from the drop-down menu at the top of the page, you can jump from one question to another or click on **End Test** to finish.

- **Timer.** If your instructor has set a time limit for the assignment, the timer will be active and it will indicate the amount of time remaining for the assignment.

- **Question text.** Be sure to read the question carefully.

- **Submit Button or Answer Field.** Depending on the type of question, the method of entering answers will differ.

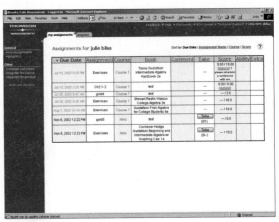

An Assignment Screen

Types of Questions

Assignments may contain questions in various, easy-to-use formats:

- **Multiple-Choice and True/False.** Click on the radio button for the correct answer.

- **Tables and Fill in the Blank.** Type the correct answer in the answer field.

- **Matching.** Two rows of items are provided; the task is to match the items on the left side with the appropriate items on the right. Click on an item at the left until the handle turns green; then use the mouse to drag the item to the appropriate item at the right. A line will link the two. If you want to change your answer, click on the **Clear All** button above the question; this will clear your choices, and you can then rematch the items.

- **Essay.** Type an extended response freely in the space provided.

- **Free Response with Math Toolbar.** Assignments for math courses have a toolbar located above the answer field with a series of tools for writing mathematical expressions.

- **Show Your Work.** Demonstrate your steps in reaching an answer.

- **Sketch.** Use drawing tools and models to sketch a solution.

- **Drag and Drop.** Drag terms to the right spot.

- **Multi-Answer.** Give short answers to several related questions.

- **Multistep/Combination.** Answer questions in a series of steps building toward a solution.

- **Animated Tables and Flash Problems.** Follow the cues in these interactive questions.

Many other kinds of questions have been designed specifically for such disciplines as math; statistics; chemistry, physics, and other sciences; philosophy; and foreign languages.

How to Enter Answers

1 In the answer field, solve question 1; then click on **Submit.**

   a. If your assignment is in Practice mode, you will be told whether your answer is right or wrong. If the answer is correct, the program will load the next question; if the answer is wrong, you can try answering again.

   b. If your assignment is in Quiz or Exam mode, you will not be told immediately whether your answer was right or wrong, only that it has been submitted. For a quiz, correct answers will be available after you have completed the assignment. Your instructor can tell you whether the assignment is in Practice, Homework, Quiz, or Exam mode.

   c. If your assignment is in Homework mode, you will be able to answer questions three times unless your instructor has specified otherwise. You will also be allowed to view any hints and solutions that are available.

2 In the question menu, click on the next question, and solve it as in step 1 above. When you are done with the test, click on **Done**; then, on the next screen, click on **End Test.** A "Result Details" chart will appear.

The "Result Details" Chart

The "Result Details" chart will appear after you have completed an assignment. The chart is a record of your assignment and includes the following important information:

• Assignment name, overall score, times taken, and time spent.

• Item, score, possible score, and right/wrong answers for each question.

Of greatest importance are details available in the "Right/Wrong" column. You should use this material to reinforce what you have done correctly and to learn from your mistakes.

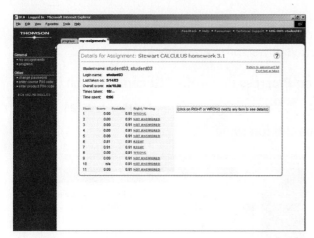

"Result Details" Chart

To review your results:

1 In the "Right/Wrong" column of the chart, click on **Right** or **Wrong** next to any item to view the original question, your answer, and the correct answer.

2 At the top right, click on **Return to Assignment List.** The "My Assignments" page will open, and you can see how your work has been recorded.

   a. In the "Take" column, your attempt is now recorded; that is, "(0/--)" has become "(1/--)."

   b. In the "Scores" column, your score is now recorded, and the **details** link appears.

You can click on the **details** link to access the "Result Details" chart at any time.

The "Progress" Area

The "Progress" page lets you check on your personal gradebook.

1 From the "Start Up" page, click on **Progress** under the "General" option in the menu at the left of the screen. A chart will open. The chart serves as a personal gradebook and contains the following:

- Course

- Assignment name

- Assigned date

- Due date and time

- Taken-on date

- Possible score

- Extra credit (if any)

- Score

- Notes from the instructor (if any)

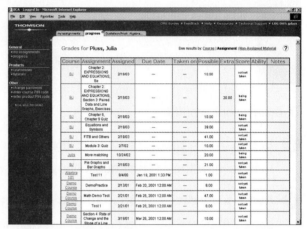

Gradebook on the "Progress" Page

2 The chart can be printed. Beneath the chart, click on **Display in separate window for printing, saving, or e-mail.**

3 In the browser's menu, click on **File**; then click on **Print.**

Changing Your Password

To change your password, click on **Change Account Information** under "Other" in the menu at the left of the screen. Change your account details, and then click on **Save Changes.**

Registering for Additional Courses and Products

Situation: You have already created an account, and now an instructor for another course gives you a new Course Access Code.

1. Click on **Enter Course Access Code** under "Other" in the menu at the left side of the screen.

2. In the **Access Code** box, enter the access code your instructor has given you; then click on **Submit.**

3. You will see the "Congratulations" screen. You will see assignments for this new course in "My Assignments."

Situation: You have already created an account, and now you have a new textbook that also uses iLrn Tutorials.

1. Click on **Enter Content Access Code** under "Other" in the menu at the left side of the screen.

2. In the **Access Code** box, enter the access code packaged with your book or the ISBN number on the back cover of your new book, then click on **Submit.**

3. You will see the "Congratulations" screen. On the menu at the left, click on **Tutorials** and begin working.

The "Tutorials" Area

Additional tutorials are provided for many books. The tutorials, keyed to the book by chapter and section, provide well-directed help for problem solving.

Accessing Tutorials for Additional Courses and Products

1 After having successfully registered to use a new product, click on **Tutorials** in the main menu at the left of the screen.

2 Click on the picture of your book to access the tutorial.

3 Choose a chapter and section from the drop-down menus below the top navigation bar. The activities to choose from vary according to the book in use, but frequently you will find the following:

- **Read Book.** PDF files taken directly from the book illustrate the concepts presented in each section, often including worked examples and defining and clarifying key concepts. Each explanation corresponds to a question or problem you are asked to complete and can be launched from within the relevant exercise.

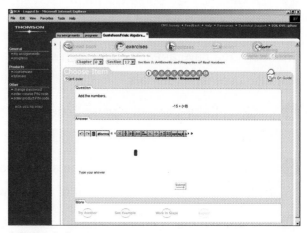

A Tutorials Screen

- **Exercises.** The exercises provide step-by-step guidance in solving the problems in each section. An unlimited number of interactive problems are available for you to work on, with options like Try Another, See Example, and Work in Steps. Queries at each step keep you directed toward a correct solution. If, by the third attempt, the correct answer has not been entered, the tutorial will display the correct answer. The Read Book option also provides a direct link to the most relevant explanation from the text.

- **Quizzes.** These provide sets of 10 questions you can use to assess your mastery. Be sure to access each item in the Item menu to complete the quiz.

- **Video.** SkillBuilder sessions use video clips to present problems and demonstrate complete solutions.

- **vMentor Tutoring.** Click to enter a vMentor classroom where live instructors assist you with problems and homework.

- **Chapter Tests.** Each chapter has a 25-question chapter test that allows you to assess understanding of the chapter's key concepts.

- **Explorations.** Multistep questions apply real-life scenarios to the concepts taught in a chapter, offering a dynamic and interesting context in which to practice your skills. These are available at the chapter level.

Other Resources

When you need help, use the links in the upper right corner:

- **FeedBack.** Use this form to let us know how iLrn is working for you.

- **Help.** Get context-sensitive help that makes sure you have the answers you need.

- **Resources.** Access glossaries and tools for your discipline.

- **Technical Support.** Go to the iLrn Technical Support Website for FAQs, student guides, information on new products, downloads, Report a Bug and feedback forms, JRE (Java Runtime Environment) instructions, and contact information for online help.

Appendix A
Using iLrn in a Mathematics Course

Math Tutorial Quick Start Guide

Note: If you already have an iLrn account, you can log in and enter new Content or Course Codes from the Main Menu. Otherwise, register according to your situation, below.

New Registration with Content Access Code

Situation: You have a textbook with an iLrn Content Access Code. You have not received a Course Access Code or password from an instructor.

1 Go to **http://www.ilrn.com**.

2 Click **Student Tutorial**.

3 Follow the on-screen instructions to select your school, if necessary.

4 In the **Access Code** box, type the Content Access Code supplied on your iLrn card. In the **ISBN** box, enter the ISBN number from your book's back cover. In the **E-mail** box, enter your valid e-mail address. Click **Submit.**

5 Follow the on-screen instructions to enter password and contact information.

6 Click **Register and Enter iLrn** to finish registration and log in.

Notes:

• You'll be using your e-mail address as your iLrn login.

• Keep your password private and in a safe place. You will need it each time you log in.

New Registration with Course Access Code

Situation: Your instructor has provided you with a Course Access Code.

1 Go to **http://www.ilrn.com**.

2 Click on **Student Tutorial**.

3 Follow the on-screen instructions to select your school, if necessary.

4 In the **Access Code** box, type the Course Access Code provided by your instructor. In the **ISBN** box, enter the ISBN number from your book's back cover. In the **E-mail** box, enter your valid e-mail address. Click **Submit**.

5 Follow the on-screen instructions to enter password and contact information.

6 Click **Register and Enter iLrn** to finish registration and log in.

Login

Once you have registered for iLrn, you can log in directly.

1 Go to **www.ilrn.com**.

2 Click **Login**.

3 Enter your iLrn login and password. Click **Login**.

Welcome to iLrn!

If you registered a product that has additional tutorials or courseware, you will see the appropriate links on the Main Menu.

Registering New Content or Courses

Once you have registered for iLrn, you can register for additional content (books or CD products) or courses easily.

1. Log in to iLrn.
2. From the **Main Menu > Other**, select **Enter Content Access Code** (to register a new book or CD) or **Enter Course Access Code** (to enter an iLrn class enrollment code). You may also see a **Self-Enrollment** option for your course.
3. Follow the on-screen instructions to enter the appropriate code.

Getting Started with Tutorials

You will see any tutorials your instructor has assigned to you on the "My Assignments" page (**Main Menu > My Assignments**).

To start an assignment, click its **Take** button. The work you complete will be noted in your instructor's gradebook.

If you registered a product with additional tutorials, you can click on the **Tutorial** link on the Main Menu. These exercises provide additional practice and exploration with the textbook content beyond that assigned by the instructor.

Select your textbook; then use the menus to select a chapter, section, and activity. These vary by book, but you will frequently find book explanations, exercises, quizzes, video tutorials, live online tutorials, chapter tests, and explorations.

If you need live tutoring while working on an Explanation, Exercise, or Quiz, click on the vMentor icon in the upper right corner of the screen.

Note: vMentor live tutoring is available Sunday through Thursday, from 8:00 P.M. to midnight Eastern time (5:00 P.M.–9:00 P.M. Pacific time). The microphone and speakers must be activated on your computer or you will need a phone line.

iLrn Can Help You Get Better Grades in Math

iLrn is an online program that facilitates math learning by providing resources and practice to help you succeed in your math course. Your instructor chose to use iLrn because it provides online opportunities for learning (Explanations found by clicking **Read Book**), practice (Exercises), and evaluating (Quizzes). It also gives you a way to keep track of your own progress and manage your assignments.

The mathematical notation in iLrn is the same as that you see in your textbooks, in class, and when using other math tools like a graphing calculator. iLrn can also help you run calculations, plot graphs, enter expressions, and grasp difficult concepts. You will encounter various problem types as you work through iLrn, all of which are designed to strengthen your skills and engage you in learning in different ways.

Navigating through iLrn

To navigate between chapters and sections, use the drop-down menu below the top navigation bar. This will give you access to the study activities available for each section.

The view of a tutorial in iLrn looks like this.

Math Toolbar

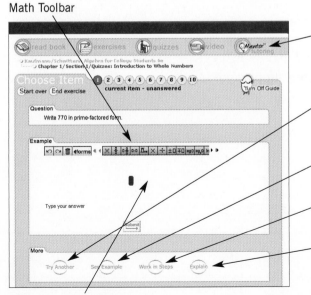

Type your answer here.

vMentor: Live online tutoring is only a click away. Tutors can take screen shots of your book and lead you through a problem with voice-over and visual aids.

Try Another: Click here to have iLrn create a new question or a new set of problems.

See Examples: Preworked examples provide you with additional help.

Work in Steps: iLrn can guide you through a problem step-by-step.

Explain: Additional explanation from your book can help you with a problem.

Online Tutoring with vMentor

Access to iLrn also means access to online tutors and support through vMentor, which provides live homework help and tutorials. To access vMentor while you are working in the Exercises or "Tutorial" areas in iLrn, click on the **vMentor Tutoring** button at the top right of the navigation bar above the problem or exercise.

Next, click on the **vMentor** button; you will be taken to a web page that lists the steps for entering a vMentor classroom. If you are a first time user of vMentor, you might need to download Java software before entering the class for the first time. You can either take an Orientation Session or log in to a vClass from the links at the bottom of the opening screen.

All vMentor Tutoring is done through a vClass, an Internet-based virtual classroom that features two-way audio, a shared whiteboard, chat, messaging, and experienced tutors.

You can access vMentor Sunday through Thursday, as follows:

> 5 p.m. to 9 p.m. Pacific time
>
> 6 p.m. to 10 p.m. Mountain time
>
> 7 p.m. to 11 p.m. Central time
>
> 8 p.m. to midnight Eastern time

If you need additional help using vMentor, you can access the Participant Guide at this website: **http://www.elluminate.com/support/guide.pdf**.

Types of Math Problems in iLrn

The problems in iLrn are algorithmically generated, which means that the program can create an infinite number of similar problems. So you can practice your math skills and work similar problems as many times as you need to master them. The following list explains the types of problems you might find while working through iLrn.

- **Sketch.** Use drawing tools and models for freehand sketching of mathematical graphs.

- **Drag and Drop.** Solve problems by using your mouse to drag the correct answer to the appropriate place on screen.

- **Animated Tables.** These problems allow you to see the principles discussed in action.

- **Multistep Problems.** iLrn breaks a problem down into steps and guides you in solving a problem one step at a time. The different questions build on each other to lead up to the final correct answer.

- **Free-Response Math.** Use the Math Toolbar above the answer field to write out mathematical expressions.

- **Multiple Math Answers.** Answering these questions requires short answers to several related questions.

- **Geometry.** These problems use shapes as correct answers. Click on the correct shape to record your answer.

- **Graph 3D.** You can modify or rotate graphs to correctly plot functions.

Answering Math Problems in iLrn

You'll use the Equation Editor to answer "free-response" homework and test questions in the correct format. The Equation Editor provides menu options for all the functions and math formatting you'll need for your answers, and evaluates your responses with sophisticated grading logic.

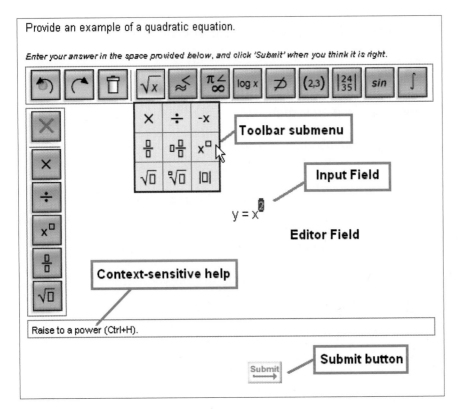

Provide an example of a quadratic equation.

Enter your answer in the space provided below, and click 'Submit' when you think it is right.

Toolbar submenu

Input Field

$y = x^2$

Editor Field

Context-sensitive help

Raise to a power (Ctrl+H).

Submit button

Equation Editor

More Math Resources in iLrn

Click on **Resources** in the menu at the top right of the screen in iLrn to access these additional tools to help you as you work in iLrn.

- **Calculator.** This online scientific calculator functions easily and intuitively, in the same way as a hand-held calculator. It includes trigonometric and logarithmic functions.

- **2D and 3D Graphers.** These visual tools allow you to enter equations and see the resulting curves and functions.

    The 2D Grapher. Enter an equation like $y = \sin(x)$, an inequality like $y > \log(x)$, or a formula for a parametric curve like $x = \ln(t)$; $y = (\cos)t$. Then just click on **Apply** and view the graph on the coordinate plane. Or use the Shapes Library to find the simple syntax for complicated curves. Overlay a second function on the first to find intersections, and customize the graph's axes.

    The 3D Grapher. This feature adds more than just a z-axis. Graph an equation; and then rotate the perspective in all three dimensions, drag the surface, set up individual drag points on the grid . . . even switch perspectives for viewing with 3D glasses.

- **Math Terms Glossary.** This list gives easy-to-understand meanings for more than 120 math terms.

- **Math Syntax Reference.** Use this reference guide for direct keyboard entry syntax.

- **Shapes Library.** This tool helps you make the connection between graphs and shapes. Use the Shapes Library to find the simple syntax to graph an arc, a conic section, a polygon, and so on. Shift, scale, mirror, rotate, or fill in any shape.

- **Units Translation.** This quick conversion calculator can change units for you. Convert angles to radians, miles to meters, even carats to tons. Types of units range from rate of change in volume to angular acceleration.

Finding Help in iLrn

Reaching technical support is easy from any screen in iLrn. Click on the **Help** button in the menu at the top at any time to be taken to an iLrn Help Guide.

To reach Technical Support directly, click on the **Technical Support** button in the menu at the top or go to the iLrn Technical Support Website by entering this address into your web browser: **http://ilrn-support.com/**. The iLrn Technical Support Website contains useful links and information, including a full list of frequently asked questions (FAQs), information on downloads, a feedback form, and a bug report form to use if you find technical problems within the program.

For more help, contact iLrn Technical Support by phone or e-mail.

Phone: 800-423-0563

Hours: Monday – Friday, 8:30 A.M. – 6:00 P.M. Eastern time

E-mail: **tl.support@thomson.com**

FAQs (Frequently Asked Questions)

More questions and answers are available at **http://ilrn-support.com/**.

❶ How do I enroll in a course?

There are two ways that a student can enroll in a course:

a. **Course Access Code.** Your instructor can create and give you a Course Access Code that allows you to automatically enroll in a course. When you register for the first time using this Course Access Code (and the URL supplied by the instructor), you will be automatically enrolled and have student-level privileges to all of the tutorial and courseware products associated with the instructor(s) of that course. For courseware, you will also need to use the Content Access Code packaged with your book.

b. **Name and password.** Your instructor may import class rosters directly into iLrn. In this case, a name and a password will be provided to you. You can enter these at the iLrn login screen. Once you have logged in to iLrn, you should change your password. Your instructor may also issue you a verification name to be used in case you lose your password.

❷ What if I don't want to enroll in a course but I want to use an iLrn product, such as a tutorial?

Your textbook comes with an iLrn insert that carries the Content Access Code for the product you have purchased. You can use this Content Access Code to gain access to iLrn products locally, on your LAN, or over the Internet.

Note: Using this Content Access Code alone does not register you into a course.

3 How do I change my password or user name?

To change your password or user name:

a. Log in to iLrn.

b. Click on **Change Account Information** on the left side of the screen. This will open up the **Change Account** tab. From this page, you can change your password, login name, and contact information.

c. Enter the appropriate information into the form.

d. Click on **Save Changes.**

4 How do I take an assignment?

Click on **My Assignments** in the main menu. A list of assignments for the course you are registered in will appear. If you are a first time user and the course has already started, there will be several assignments listed by date.

5 If I do not complete an assignment, can I go back to finish it later?

All assignments are designed to be suitable for a single sitting, so there is no save-and-resume function. Assignments can be set up to be taken multiple times, and you can take any length of time to complete an assignment.

6 If my computer crashes while I'm working on an assignment, is my work lost?

The database that provides iLrn its foundation saves material in real time so most material is saved in the event of a power or connection disruption. But in some situations, if power is lost at just the right moment during a client-server transaction, some data loss may occur.

If you experience a computer error during a web-based assignment, you have up to 15 minutes to resume taking that assignment at the point where you left off.

Getting Downloads to Use iLrn

You will need to download certain programs onto your computer in order to run iLrn properly. These are common programs in wide use, and most might already be on your computer.

See the section on downloads at the following website:

http://ilrn-support.com/

Using Other Brooks/Cole Math Technology Learning Tools

Interactive Video Skillbuilder CD-ROM or Digital Video Companion

The free CD for this program is bound into the back of your book. It contains hours of video instruction, tutorial problems, and quizzes. The following features are included in each section of the CD-ROM. They can be accessed from the main menu.

* **Video Lesson.** Video Lessons allow you to view and hear worked-out examples covering key topics in the section. Use the round buttons at the bottom of the screen to pause, start, or rewind, and the scrollbar beneath the video to move quickly back and forth through the video.

* **SkillBuilder.** SkillBuilder sessions present problems, score your answers, and provide complete solutions with step-by-step explanations to each problem. Score reports can be printed for your records or to hand in for credit or extra credit. Use the buttons at the bottom of your screen to navigate. To answer multiple-choice problems, click on the desired answer. For open-entry answers, enter the answer in the box provided and click on **Submit.**

- **Section Quiz.** Section Quizzes provide algorithmically generated problems and give immediate feedback when you enter your answers. You can work default problem sets or create customized quizzes that include problems from one or more sections and/or chapters. To work a preselected default mix of problems, click on **Begin Working Problems.** If you wish to change the number of problems in the quiz, enter the desired number and click on **Update** before beginning your session.

To create a customized quiz to include problems from one or more sections and/or chapters, click on **Custom Quiz** and follow the on-screen instructions. After you select a chapter and section, sample problems will be displayed and you can specify how many of each type of problem you wish to include in your quiz. Click on the **Update** button to update your choices. Click on **Begin Working Problems** when you're ready to start the quiz. Score reports can be printed. Use the buttons at the bottom of your screen to navigate. To answer multiple-choice problems, click on the desired answer. For open-entry answers, enter the answer in the box provided and click on **Submit.**

Note: Macintosh users need an active Internet connection to access SkillBuilders and Section Quizzes. Click on the **Web Quiz** button in the upper right corner of your screen to connect to the iLrn Internet–based testing, tutorial, and course management system. Help for iLrn is available on the site.

- **Solution Finder.** While working SkillBuilder problems, you can access Solution Finder. This tool allows you to enter your own basic problems and receive step-by-step help, as you would from a tutor. Click on the **Solution Finder** button in the upper right corner of your screen to access this feature.
  Note: This button is only visible while you are working in a SkillBuilder session.

- **Chapter Test.** After choosing a chapter from the main menu, you can select the Chapter Test. This test consists of algorithmically generated problems with feedback given only after the session is completed.

Follow these instructions to log in and navigate through the Skillbuilder CD-ROM.

Login

1 Enter your first and last names. Other information is optional but may be appropriate if you plan to hand in a printed score report for credit or extra credit. This score report includes all the information you enter at login.

2 When you have entered the desired information, click on **Log in.**

Main Menu

1 Click on a chapter title to reveal all the available sections and the test for that chapter.

2 Click on a section title to view the Video Lesson, SkillBuilder, or Section Quiz for that section, or click on the Chapter Test to test your knowledge of the chapter topics.

3 Click on the **Main Menu** button in the upper right corner of your screen to return to this menu from anywhere in the program.

Quit

To quit the program, click on the **X** in the upper right corner of your screen.

The Learning Equation

The Learning Equation (TLE) is a multimedia courseware and workbook package developed as an online learning tool for developmental mathematics. If your instructor has adopted TLE, do the following.

1 Log in through iLrn; then click on **Courseware** in the navigation bar at the left.

2 Click on the TLE cover to enter the TLE course.

TLE is designed to engage you in your own learning while building skills in algebra and problem solving. As you progress through the lessons, you will practice key concepts, build your vocabulary, and develop your basic math and reasoning skills.

A complete "User's Guide for TLE" can be found at the beginning of your TLE workbook. Technical support is the same for TLE and iLrn. Refer to the section titled "Finding Help in iLrn" in this appendix for more information.

Other Courseware Books

Other interactive electronic books from Brooks/Cole can be accessed by clicking on **Courseware.**

Appendix B
Using iLrn for Statistics: iLrn Homework with DuxStat

Welcome to iLrn Homework with DuxStat

iLrn's online environment allows you to monitor your own progress and turn in your homework and exams online. And with exercises with data sets from selected textbooks integrated right into iLrn, you can focus on interpreting statistical output—because DuxStat *computes all the calculations for you.* In a typical exercise, such as the problem shown below, you can perform a graphical analysis and an inferential analysis and then interpret the results.

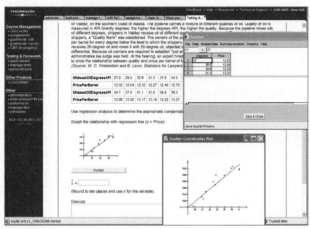

Graphical Analysis in DuxStat

DuxStat is fully integrated into your homework and assessment problems and offers you immediate feedback on your work. Developed by a professor of statistical computing, Dr. James Hardin of Texas A&M University, the DuxStat Java-based statistical analysis applet covers all tests and graphs needed for introductory courses.

Benefits of iLrn with DuxStat

- iLrn with DuxStat is easy to use. With one click, DuxStat opens with the right data loaded for the problem. You can then select the appropriate action, and the results of your analysis are automatically transferred into the answer space.

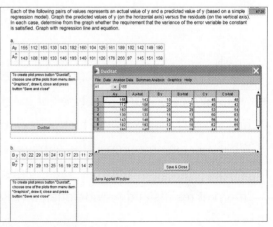

DuxStat Opens with Right Data Loaded

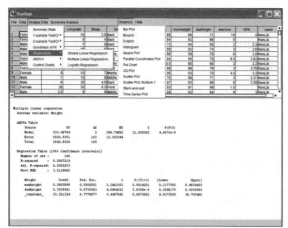

Data Analysis Software

• Using iLrn with DuxStat, you can focus on data analysis and higher-level thinking, rather than number crunching. DuxStat problem types have a data analysis tool that you can use *within* the homework and testing environment. You can also use other data analysis software and enter the results from those calculations online.

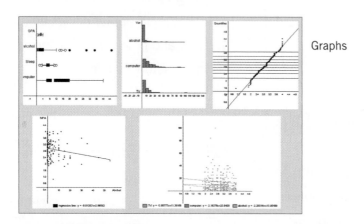

Graphs

• DuxStat creates numerous types of graphs for you, including scatterplots, histograms, stem-and-leaf plots, dotplots, QQ plots, boxplots, and many more. When you create a graph, it is automatically loaded into the homework or test problem, and it appears onscreen immediately. You can then use that graph to interpret results and provide qualitative analysis of the data.

Appendix C
Troubleshooting Guide: Java-Related Problems

Having the correct version of Java installed on your computer is essential to being able to use iLrn. If you are experiencing any of the following problems, you may have a problem with Java on your computer:

- The browser freezes at "Testing Browser."

- You get an internal error when taking a test.

- Graphs do not load properly.

The following information will help you find out if Java is the problem in Internet Explorer for Windows.

1 Check to see if Java is working correctly:

a. Open Internet Explorer.

b. Click on **Tools** and then **Internet Options.**

Internet Options on Menu

c. Click on the **Settings** button.

Settings Button

d. Click on the **View Objects** button.

View Objects Button

If the status of the Java Runtime Environment (JRE) is listed as damaged or the total size is listed as "None," as in the following example, you need to remove the damaged Java object.

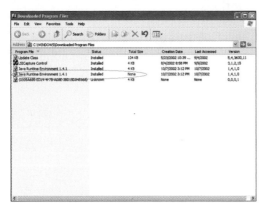

Java Runtime Environment (JRE)

e. To remove a damaged Java object, right-click on the Java object and then click on **Remove.** You may get an error message that says there is not enough information to remove Java. Click on **OK** to continue.

f. Close Internet Explorer; then open it up again. If this procedure doesn't resolve the problem, you will need to get a new version of Java.

2 Check to see if you have Java installed:

a. Double-click on the **My Computer** icon. The My Computer window will open.

My Computer Icon

b. Double-click on **Control Panel.**
The Control Panel window will open.

c. Double-click on **Add/Remove Programs.**
The Add/Remove Programs window
will open.

d. Look for Java 2 Runtime Environment;
if you see multiple copies, remove
the oldest version. In the previous
example (in c above), you would
remove the first one, Java 2 Runtime
Environment, SE v1.4.0_02.

3 If Internet Explorer shows that Java is
installed and working correctly, the
problem may be that you have an older
version of Java. The solution then is to
get a new version of Java:

a. Go to **http://www.java.com/en/download/
windows_automatic.jsp.** Java will be
downloaded and installed automatically.

b. During download and installation, a series of windows will appear: Click on
Yes or **Next** on each to continue. Download times vary: 15–30 minutes is
typical for a 56K modem. Once the download is complete, installation takes
only about three minutes.

You're ready to go!

Control Panel Icon in
My Computer Window

Add/Remove Programs Icon in
Control Panel Window

Appendix D
iLrn Problem Types

Standard Problem Types

These are among the 50 problem types found in iLrn.

- **Multiple-Choice and True/False.** Simply click on the radio button for the correct answer.

- **Fill in the Blank.** Type the correct free-response answer in the answer field.

- **Matching.** Two rows of items are provided; the task is to match the items on the left side with the appropriate items on the right side. Click on an item at the left until a blue box appears around it; then, holding down the button, drag the mouse to the appropriate item at the right. An arrow will link the two. If you want to change your answer, click on the garbage can button above the question; this will clear your choices, and you can then rematch the items. In a second type of matching problem, letters attached to items in left column are dragged to blanks attached to items at the right.

- **Tables.** Type the correct answer into blank cells in the table.

- **Narrative.** Answer several questions based on one block of text. These problems are similar to SAT reading comprehension questions. Narrative problems may also include a table of values for scientific parameters, and so on. Narrative is a general property of every problem.

- **Essay.** Type an extended response.

Unique, Visual, and Interactive Problem Types

These different problem types are designed to more effectively assess your real understanding of concepts:

- **Flash-Based Problems.** Integrating sound, motion, and interactivity, these problem types help you learn in a completely new way.

- **Sketch.** These problems provide drawing tools and models for sketching solutions.

- **Drag and Drop.** You drag and drop terminology to demonstrate your mastery of concepts.

- **Animated Tables.** These problems allow you to see principles in action.

- **Multistep/Combination.** The solution is not just a simple response; these problems require you to answer a series of questions (of different types) that build on one another.

- **Multiple Math Answer.** In these problems, several related questions are presented, each requiring short answers.

Specialized Problem Types for Specific Course Areas

iLrn offers a wide variety of problem types for courses in statistics and chemistry.

Statistics

- **Spreadsheet Problems.** Test your understanding of a basic statistical concept and use a spreadsheet to calculate the results.

- **DuxStat Problems.** These problems allow you to import data sets, calculate statistics on them using more than 20 methods, build advanced models, and display statistical graphs. DuxStat's spreadsheet-like Data Editor can load data from textbooks or data files from the web and allows you to create your own data sets.

Chemistry

- **Chemical Bond/Node Choice.** You can identify double bonds in these 2D visual problems.

- **ChemConstructor/Molecule Builder.** You are provided with the tools that allow you to build 2D representations of simple or complex chemical molecules from a palette of molecule fragments.

- **Functional Group.** You can mark chemical functional groups and name them.

- **Lewis Structure.** You can build a Lewis structure for the given chemical formula and/or build balanced equations of reactions using the Lewis structure of reagents and products.

- **Resonance.** You can build chemical formulae with unshared electrons shown and ionization designated by special characters.

- **Stereo Isomers.** You can give multiple answers in a form of stereo projection.

- **Energy Levels.** You can place electrons in available bins to model the correct configuration in response to a question about electronic energy states in atoms or molecules.

Survey/Feedback Form for iLrn

Thank you for using iLrn. We hope iLrn is the easiest and most powerful learning system you have ever experienced. But we want to keep on making it better. Please let us know what worked—and didn't work—for you.

Name (optional): _____

Discipline: _____

Course: _____

Instructor's name: _____

School: _____

1 How often did you use iLrn?
❏ Every day
❏ Three to five days a week
❏ Once or twice a week
❏ Occasionally

2 How easy was it for you to log in to iLrn?
❏ Very easy
❏ Easy
❏ Difficult
❏ Extremely difficult

3 Did you use iLrn as part of a course set up by your instructor?
❏ Yes
❏ No

4 If yes, how would you rate the course management features of iLrn? (Rate from 1 to 4, with 1 as the best; select "Not used" for any features you did not use.)

Ability to see my grades	1 2 3 4	Not used
Ability to access assignments online	1 2 3 4	Not used
Ability to e-mail my instructor from within the program	1 2 3 4	Not used
Ability to receive notes and comments from my instructor	1 2 3 4	Not used

5 If you used the iLrn Tutorials in Mathematics, how do you rate the following features within iLrn? (Rate from 1 to 7, with 1 as the best; select "Not used" for any features you did not use.)

Ability to try as many problems as I want	1 2 3 4 5 6 7 Not used
Ability to work in steps	1 2 3 4 5 6 7 Not used
Ability to get feedback on my work	1 2 3 4 5 6 7 Not used
Ability to access skillbuilder video explanations	1 2 3 4 5 6 7 Not used
Ability to access online live tutoring through vMentor	1 2 3 4 5 6 7 Not used
Ability to see the pages of my textbook with explanations	1 2 3 4 5 6 7 Not used
Ability to test myself through quizzing	1 2 3 4 5 6 7 Not used

6 If you used iLrn with DuxStat, how would you rate the DuxStat features? (Rate from 1 to 6, with 1 as the best; select "Not used" for any features you did not use.)

Ability to have calculations done for me	1 2 3 4 5 6 Not used
Ability to have the results of my analysis automatically transferred into the answer space	1 2 3 4 5 6 Not used
Ability to receive feedback on my work	1 2 3 4 5 6 Not used
Ability to easily create graphs and have the graphs appear on-screen immediately	1 2 3 4 5 6 Not used
Ability to e-mail my instructor from within the program	1 2 3 4 5 6 Not used
Ability to receive notes and comments from my instructor	1 2 3 4 5 6 Not used

7 What capabilities would you like to see in iLrn?

_____

8 What should be improved in iLrn?

_____

Thank you for your feedback. Please return to **iLrn Survey**
c/o Margaret Parks • Brooks/Cole • 10 Davis Drive • Belmont, CA 94002